To my mother-in-law, Sue,
a retired librarian whose favorite place
is in the pool with a good book.
Thank you for all the years
of encouragement.

Beach Lane Books
New York   Amsterdam/Antwerp   London   Toronto
Sydney/Melbourne   New Delhi

Welcome to a rainforest celebration!

Hear rain showers tickle the treetops!

See rainbows color the canopy!

Watch raindrops
dance down into the forest...

Soon, algae and microbes made an appearance. They dug into the party platters of decomposing leaves.

Meanwhile, the bromeliad brewed up a pool-party punch of leaves, seeds, and sludge.

As even rainier days set in,
the party became a lively kiddie pool.

After many wet months, the rain showers slowed and the canopy dried out.

But the sky-high pool party stayed in full swing, booming with a chorus of frogs, flies, and crickets.

Parrots partied down.

Protozoan popped up.

"Did we miss anything?"

And damselflies fluttered to toucan tunes.

"They're singing my favorite song!"

Croak!

Crick!

Lion tamarins whistled.
Manakins whirred.
And howler monkeys whoop-whooped!

Snails scrounged on leftovers.

Millipedes mingled in moss.

"I call dibs on this moldy leaf."

"Hi, Midge!"

"Hi, Milli!"

And a woodnymph took home party favors.

The pups grew and grew and stretched their young leafy rings open, welcoming...

rain showers that tickle the treetops.

Rainbows that color the canopy.

Raindrops that dance down to form freshwater pools among the budding bromeliads.

More about Bromeliad Pools

Bromeliads play an important role in rainforest diversity. The small rainwater pools that form inside their leaf structures provide food, water, and shelter for a variety of tree-dwelling species throughout the year. The pools collect seeds and debris that rainforest critters eat, and they are also a water source up in the canopy. Bromeliad flowers, which bloom only once in the plant's life, provide nectar and pollen for hummingbirds, bats, bees, and butterflies. Some critters, like the yellow heart-tongued frog, live their entire life cycle in bromeliad pools. And bromeliad insect and amphibian nurseries attract large predators, like primates and birds.

Bromeliads are native to the Neotropics. Some bromeliads live on the ground while others, like the one in this book, are epiphytes, meaning that they grow on other plants. These bromeliads obtain all their nourishment by absorbing what they need from their pools and open-air roots and are not harmful to their host trees. In fact, some bromeliads are beneficial, providing homes for ants that fend off tree-boring beetles.

This story takes place in the Atlantic Forest in South America over the course of a year. The forest experiences two seasons a year: a dry season from about April through September and a rainy season from about October through March. Our bromeliad, *Aechmea victoriana*, is perched alongside orchids in a native *Miconia hypoleuca* tree that blooms in the rainy season and fruits in the dry season. Golden lion tamarins, along with other tree-dwelling critters, eat this tree's fruit and seeds in addition to bromeliad leaves throughout the year. The bromeliad blooms in the dry season in this region and is visited by bees and hummingbirds from July through September.

The Atlantic Forest is South America's second-largest rainforest after the Amazon. It was once almost twice the size of Texas. But after centuries of deforestation, it is now less than 7 percent of its original size. It is home to 23,000 species of plants and more than 2,200 species of birds, mammals, reptiles, amphibians, and fish, many of which are endangered. Conservationists and local communities are now working to restore this forest habitat, and bromeliads play a part in helping to sustain the rare and wonderful species that live here.